Pioneers of Science in America

Pioneers of Science in America

Collection

LM Publishers

Benjamin Franklin[1]

We are here to unveil memorial busts of Americans distinguished in science. I am honored by the privilege of speaking of Benjamin Franklin. This man, the father of American Science, was possessed of mental gifts unequaled in his day. Even yet he holds the highest place in the intellectual peerage of a land where, in his time, men had few interests which were not material or political. But no man entirely escapes the despotic influences of his period. Thus in every life there are unfulfilled possibilities, and so it was that, paraphrasing Goldsmith, we may say that Franklin to country gave up what was meant for mankind, when with ~~deep regret he resigned, in~~ middle life, all

[1] By Dr S. Weir Mitchell

hope of whole-souled devotion to science. When most productive his scientific fertility was the more remarkable because of the other forms of dutiful activity which in a life that knew no rest left small leisure for those hours of quiet thought without which science is unfruitful of result.

There is a Hall of Fame not built by the hand of man. It is the memory of mankind. In many of its galleries this man's bust could with justice be placed. Diplomacy would claim him as of her greatest. For him would be the laurel of administrative wisdom. Among states- men he would be welcomed; and who of the masters of English prose shall in that hall of fame be more secure of grateful remembrance, and who more certain of a place among men of science.

As an investigator of nature and of nature's laws he is materially represented here by right of eminent achievement. Let us as men of science feel proud that Franklin's fame as a philosopher did much to win for Franklin the diplomatist such useful consideration and respect as led to final success.

Many of those you honor today had moral and temperamental peculiarities which more or less influenced their lives and are common to men of science. Most of them cared little about making money; still less about keeping it.

Franklin, on the contrary, dreaded poverty; was careful in business, made fruitful investments and died rich; nevertheless, like the typical man of science, he refused to make money out of his discoveries, or by patents to protect his inventions. In him the man of science, unselfish, free from money greed, seemed to exist apart from all those other men who went to the making of the many- minded Franklin.

In another way he was singularly unlike such typical men of science as Henry, in physics, and Leidy, in natural history. When Franklin made a discovery his next thought was as to what practical use it could be put. If he made some novel observation of nature, he asked himself at once how he could make it serve his fellow men. The great reapers of the harvest of truth commonly leave the inventor to make practical use of their unregarded thought.

Leaving the wide land to do justice to Franklin, the model citizen and great diplomatist, here we crown him with the assured verdict of posterity Franklin, the man of pure science. Here we welcome him to this goodly fellowship of those who communed with nature and read the secrets of the Almighty Maker.

Alexander von Humboldt[2]

In this immortal man, whose bust you have gathered to unveil, the world reveres its greatest master since the days of Aristotle. His genius covered all that man ever thought, did and observed in nature. There is no branch of human knowledge into which his mind did not penetrate. His Cosmos, that marvelous monument of meditation and research, is a new book of Genesis in which the universe mirrors itself in all its vastness and minuteness ' from the nebulae of the stars ' — to use his own words — 'to the geographical distribution of mosses on granite rocks.'

---

[2] By Baron S. Von Sternburg

By his wonderful talent of research, by his almost superhuman power to divine eternal laws, this great interpreter of science taught mankind how to read in the book of nature, how to understand its great mysteries. The series of sciences, originated by this mighty genius is, as well as the other manifold branches of science developed by him, sufficiently known to all of you.

In all his investigations his ultimate aim was to bring theory into practical relation with life. Thus he not only elevated the standard of culture of the whole world by many steps, but he also became from a practical point of view the benefactor of mankind in many branches of common life, as trade, commerce, navigation.

He taught us how to conceive the beauty and sublimity of nature in its every form and motion. His studies are not a matter merely of memory and of dry meditation, to him nature was rather the inexhaustible source of pure and deep enjoyment, by which the heart is purified and ennobled and men are brought nearer to perfection.

It is not necessary to give you a more detailed picture of his life. All this is so well known and so dear to the whole learned world of America; for never has a foreign scholar been more honored in this country than Alexander von Humboldt.

We need only recall the celebrations which took place in his memory, both at the time of his death and on occasion of the centennial anniversary of his birth, when throughout all America solemn offerings of gratitude and devotion went out to the shadow of the great dead.

Humboldt devoted five years of his life to scientific investigations in South and Central America, in Mexico and in Cuba. He ascertained the course of the greatest rivers, he climbed the summits of mountains, where never man's foot had trod before, he studied vegetation, astronomical and meteorological phenomena, gathered specimens of all natural products and a great deal of historical information about the early population of these parts of the New World. It was he that drew the first accurate maps of these regions. With almost prophetic forecast of the needs of generations to come, he examined the Isthmus of Panama and considered carefully the possibilities of establishing an interoceanic waterway.

It is well known how great an interest Alexander von Humboldt has taken in the United States. Indeed, so strongly was he attracted by the problems of the new-born republic that, putting aside even his habitual scientific occupations, he devoted himself entirely for some months to the study of the American people and the institutions of this country.

Finally, the great scientist, he whom people call the scientific discoverer of America, returned to his country, carrying with him a vast store of intellectual and material treasures of science. So abundant were the results reaped from his expeditions that he needed the cooperation of the best scholars of his time to compile that great mass of material, and to place it into proper shape and form.

Throughout his long and industrious life, Alexander von Humboldt has ever retained his love and devotion for the country where his great field of labor lay, and for its people to whom he always felt so closely connected by his love for freedom in thoughts and for liberty. It is a well-known fact that in his later days of all foreign people who ever knocked at his door no one was more heartily welcomed than the American citizen.

The benefits of his investigations in America returned to that country in the course of time. No wonder that her people recognize him as their benefactor. Another great man, whose monument will be unveiled today, and most deservedly placed beside the one of Alexander von Humboldt, Louis Agassiz, says of him: "To what degree we Americans are indebted to von Humboldt, no one knows who is not familiar with the history of learning and education in this country. All the fundamental facts of popular education in physical science beyond the merest elementary instruction, we owe to him," and at another place : "Let us rejoice together that Humboldt's name will permanently be connected with education and learning in this country, for the prospects and institutions of which he felt so deep and so affectionate a sympathy."

Of all the tributes that have been paid to Alexander von Humboldt the latest and most fitting has now found its expression in this building. For here, in this magnificent Museum of Natural History, the ideal aim of all his theories is realized most perfectly: to cultivate the love of nature, and thus to ennoble man and beautify his life.

Gentlemen, permit me to thank you for the honor you have clone me to-day, and to express the hope that this splendid building may become a shrine of pilgrimage for scientists and students also of the Old World, helping to bind the nations closer together.

John Torrey

John Torrey[3]

As a pioneer of American botany, John Torrey naturally finds a place among the men whose works we gladly celebrate to-day, in this grand institution, developed in the city where he was born, where he resided the greater part of his life, and where he died. To-day's recognition of Torrey as a master of botanical science is, therefore, peculiarly appropriate in New York, where he is already commemorated by the society which bears his name, by the professorship in Columbia University named in his honor, and by ids botanical collections and library deposited by Columbia University at the New York Botanical Garden.

---

[3] By Dr. N. L. Britton

Dr. Torrey was born on August 15, 1796, and died March 10, 1873, nearly thirty-four years ago; the pleasure of his personal acquaintance is, therefore, known to but few persons now living; we have abundant evidence, however, that he was honored and beloved to a degree experienced by but few; righteousness was instinctive in him, aid to others was his pleasure, he was tolerant and progressive, and his genial presence was a delight to his associates.

He was educated for the profession of medicine, graduating from the College of Physicians and Surgeons in 1818, but soon abandoned it and in 1824 became professor of chemistry at West Point; after three years service there, he was elected professor of chemistry and botany in the College of Physicians and Surgeons, a position which he held for nearly thirty years, during part of this period lecturing on chemistry also at Princeton ; he was also United States assayer in New York from 1854 until his death.

Dr. Torrey's attention was directed to botany during his youthful association with Professor Amos Eaton, and his interest in that science subsequently stimulated during his medical studies by the lectures of Professor David Hosack. It early became his favorite study, and, not-withstanding his noteworthy services to chemistry, his fame rests on his botanical researches, although they were accomplished during his hours of rest and recreation, and largely during the night.

His botanical publications began in 1819 with 'A Catalogue of Plants Growing Spontaneously within Thirty Miles of the

City of New York,' published by the Lyceum of Natural History, now the New York Academy of Sciences, and were completed the year after his death in the 'Phanerogamia of Pacific North America,' in Vol. 17 of the Report of the United States Exploring Expedition. His contributions to botany include over forty titles, many of them volumes requiring years of patient study; they throw a flood of light on the plants of North America, and form a grand contribution to knowledge. His collections, on which these researches arc based, were annotated and arranged by him with scrupulous care and exactness, and are treasured as among the most important of all scientific material in America.

Joseph Henry

Joseph Henry[4]

This time, one hundred years ago, Joseph Henry, whose name and fame we honor today, was a lad seven years of age. He was born at Albany, New York, of Scotch parentage, his grandparents on both sides having come from Scotland in the same ship to the Colony of New York, in 1775.

Doubtless he had himself in mind when in his mature years he affirmed that "The future character of a child, and that of a man also, is in most cases formed probably before the age of seven years." At any rate, he found himself early, for at the age of sixteen he had determined to devote his life to the acquisition of knowledge. Thus be became, in turn, student, teacher, civil

[4] By Dr. Robert S. Woodward

engineer in the service of his native state, professor of mathematics and natural philosophy in the Albany Academy, professor of natural philosophy in the College of New Jersey — now Princeton University — and a pioneer investigator and discoverer of the first order before he was thirty-three years of age.

His inventions and discoveries in electromagnetism especially art of prime importance. They include the inventions of the electromagnetic telegraph and the electromagnetic engine, and the discovery of many of the recondite facts and principles of electromagnetic science.

From the age of thirty-three, when he took up the work of his professorship at Princeton, till the age of forty-seven, when he was called to the post of secretary of the

Smithsonian Institution, he pursued his original investigations with untiring zeal and with consummate experimental skill and philosophic insight. It was during this period that Henry and Faraday laid the foundations for the recent wonderful developments of electromagnetic science. The breadth as well as the depth of Henry's learning is indicated by the fact that he found time during this busy period for excursions and for lectures in the fields of architecture, astronomy, chemistry, geology, meteorology and mineralogy, in addition to his lectures and researches in physics.

He was a man rich in experience and ripe in knowledge when, in 1846, he assumed the administrative duties implied by the bequest of James Smithson. "To found at Washington, under the name of the Smithsonian Institution an Establishment for the increase and diffusion of knowledge among men."

Henceforth, for thirty-two years, until his death in 1878, he devoted his life to the public service, not alone of our own country, but of the entire civilized world. In this work be manifested the same creative capacity that had distinguished his earlier career in the domain of natural philosophy. He became an organizer and a leader of men. To his wise foresight we owe not only the beneficent achievements of the Smithsonian Institution itself, but also, in large degree, the correspondingly beneficent achievements of the Naval Observatory, the Coast and Geodetic Survey, the Weather Bureau, the Geological Survey, the Bureau of Fisheries and the Bureau of American Ethnology; for to Henry, more than to any other man, must be attributed the rise and the growth in America of the present public appreciation of the scientific work carried on by governmental aid.

We may lament, with John Tyndall, that so brilliant an investigator and discoverer as Henry should have been sacrificed to become so able an administrator. And American devotees to mathematico-physical science may be pardoned for entertaining an elegaic regret that Henry as a pioneer in the fields of electromagnetism did not have the aid of a penetrating mathematical genius, as Faraday had his Maxwell. But posterity, just in its estimates towards all the world, will recognize in Henry, as we have recognized in our earlier hero, Benjamin Franklin, a many-sided man — a profound student of nature; a teacher whose moral and intellectual presence pointed straight to the goal of truth; an inventor who dedicated his inventions immediately to the public good; a discoverer of the permanent law T s which reign in the sphinx-like realm of physical phenomena; an administrator and organizer of large enterprises which have yielded a rich fruitage for the enlightenment and for the melioration of mankind; a leader of men devoted to the progress of science; a patriot, friend and counselor of Abraham Lincoln in the darker clays of the republic — in short, an exemplar for his race, a man whose purity and nobility are here fitly symbolized in enduring marble for our instruction and guidance and for the instruction and the guidance of our successors in the centuries to come.

John James Audubon[5]

Of the naturalists of America no one stands out in more picturesque relief than Audubon, and no name is clearer than his to the hearts of the American people.

Born at an opportune time, Audubon undertook and accomplished one of the most gigantic tasks that has ever fallen to the lot of one man to perform. Although for years diverted from the path nature intended him to follow, and tortured by half-hearted attempts at a commercial life, against which his restive spirit rebelled, he finally, by the force of his own will, broke loose from his bondage and devoted the remainder of his days to the grand work that has made his memory immortal.

---

[5] By Dr. C. Hart Mehriam

His principal contributions to science are his magnificent series of illustrated volumes on the birds and quadrupeds of North America, his Synopsis of Birds and the Journals of his expeditions to Labrador and to the Missouri and Yellowstone rivers.

The preparation and publication of his elephant folio atlases of life-size colored plates of birds, begun in 1827 and completed in 1838, with the accompanying volumes of text (the 'Ornithological Biography,' 1831-1839), was a colossal task. But no sooner was it accomplished than an equally sumptuous work on the mammals was under-taken, and, with the assistance of Bachman, likewise carried to a successful termination. For more than three quarters of a century the splendid paintings which adorn these works, and which for spirit and vigor are still unsurpassed, have been the admiration of the world.

In addition to his more pretentious works, Audubon wrote a number of minor articles and papers and left a series of Journals, since published by his granddaughter, Miss Maria B. Audubon. The Journals are full to overflowing with observations of value to the naturalist, and, along with the entertaining : 'Episodes,' throw a flood of light on contemporary customs and events — and incidentally are by no means to be lost sight of by the historian.

In searching for material for his books, Audubon traveled thousands of miles afoot in various parts of the eastern states, from Maine to Louisiana; he also visited Texas, Florida and Canada, crossed the ocean a number of times, and conducted expeditions to far-away Labrador and the then remote Missouri and Yellowstone Rivers. When we remember the limited facilities for travel in his clay — the scarcity of railroads, steamboats and other conveniences — we are better prepared to appreciate the zeal, determination and energy necessary to accomplish his self-imposed task.

That it was possible for one man to do so much excellent field work, to write so many meritorious volumes, and to paint such a multitude of remarkable pictures must be attributed in no small part to his rare physical strength — for do not intellectual and physical vigor usually go hand in hand and beget power of achievement? Audubon was noted for these qualities. As a worker he was rapid, absorbed and ardent; he began at daylight and labored continuously till night, averaging fourteen hours a day, and, it is said, allowed only four hours for sleep.

In American ornithology, in which he holds so illustrious a place, it was not his privilege to be in the strict sense a pioneer, for before him were Vieillot, Wilson and Bonaparte; and contemporaneous with him were Richardson, Xuttall, Maximilian, Prince of Wied, and a score of lesser and younger lights — some of whom were destined to shine in the near future.

Audubon was no closet naturalist — the technicalities of the profession he left to other — but as a field naturalist he was at his best and had few equals. He was a born woodsman, a lover of wild nature in the fullest sense, a keen observer, an accurate recorder, and, in addition, possessed the rare gift of instilling into his writings the freshness of nature and the vivacity and enthusiasm of his own personality.

His influence was not confined to devotees of the natural sciences, for in his writings and paintings, and in his personal contact with men of affairs, both in this country and abroad, he exhaled the freshness, the vigor, the spirit of freedom and progress of America — and who shall attempt to measure the value of this influence to our young republic ?

Audubon's preeminence is due, not alone to his skill as a painter of birds and mammals, nor to the magnitude of his contributions to science, but also to the charm and genius of his personality — a personality that profoundly impressed his contemporaries, and which, by means of his biographies and journals, it is still our privilege to enjoy. His was a type now rarely met — combining the grace and culture of the Frenchman with the candor, patience, and earnestness of purpose of the American. There was about him a certain poetic picturesqueness and a rare charm of manner that drew people to him and enlisted them in his work. His friend, Dr. Bachman, of Charles- ton, tells us that it was considered a privilege to give to Audubon what no one else could buy. His personal qualities and characteristics appear in some of his minor papers— notably the essays entitled 'Episodes.' These serve to reveal, perhaps better than his more formal writings, the keenness of his insight, the kindness of his heart, the poetry of his nature, the power of his imagination, and the vigor and versatility of his intellect.

Louis Agassiz

Louis Agassiz[6]

I think that the first time when I ever saw Agassiz was at one of his own lectures early in his American life. This was a description of his ascent of the Jungfrau. I think it was wholly extempore and though he was new in his knowledge of English, it was idiomatic and thoroughly intelligible.

At the end, as he described the last climb, hand and foot, by which as it seems, men come to the little triangular plane, only three feet across, which makes the summit, he quickened our enthusiasm by describing the physical struggle by which he lifted himself so that he could stand on this little three-foot table: He said, 'one by one we stood

---

[6] By the Rev. Edward E. Hale

there, and looked down into Swisserland.'
He bowed and retired.

I know I said at once that Mr. Lowell, of
our Lowell Institute, who had imported
Agassiz' (that is James Lowell's phrase),
might have said before the audience left the
hall, ' You will see, ladies and gentlemen,
that we are able to present to you the finest
specimen yet discovered of the genus homo
of the species intelligens.'

And looking back half a century, on those
very first years of his life in America, I
think it is fair to say that wherever he went
he awakened that sort of personal
enthusiasm. And he went everywhere. He
was made a professor in Harvard College in
1848. But he never thought of confining
himself to any conventional theory of a
college professor's work. He was not in the

least afraid of making science popular. He flung himself into any or every enterprise by which he could quicken the life of the common schools, and in forty different ways he created a new class of men and women. Naturalists showed themselves on the right hand and on the left. I have seen him address an audience of five hundred people, not twenty of whom when they entered the hall thought they had anything to do with the study of nature. And when after his address they left the hall, all of the five hundred were determined to keep their eyes open and to study nature as she is. From that year 1848, you may trace a steady advance in nature study in the New England schools.

That is to say, that his distinction is that of an educator quite as much as it is that of a naturalist. In 1888, Lowell said, in his quarter-millenial address at Harvard College, that the college trained no great educator, ' for we imported Agassiz.' A great educator he truly was.

When Agassiz was appointed professor he was forty-one years old. In my first personal conversation with him he told me a story, which may not have got into print, of his own physical strength. He spoke as if it were then an old experience to him. Whether he were twenty- five or thirty-five when it happened, it shows how admirable was his training and his physical constitution. He had been with a party of friends somewhere in eastern Switzerland. They were traveling in their carriages; he was on foot. They parted with the understanding that they were to meet in the Tyrol, at the city of Innsbruck.

Accordingly the next morning, Agassiz rose early and started through the mountains by this valley and that, as the compass might direct or his previous knowledge of the region. He did not mean to stop for study and they did not. But he had no special plan as to which hamlet or cottage should cover him at night. Before sundown he came in sight of a larger town than he expected to see, in the distance, and calling a mountaineer, he asked him what that place was. The man said it was Innsbuck. Agassiz said that that could not be so. The man replied with a jeer that he had lived there twenty years, and had always been told that that was the name of the place, but he supposed Agassiz knew better than he did. Accordingly Agassiz determined that he would sleep there and did so. The distance was somewhere near seventy miles. I know it gave me the impression of a walk through the valley passes at the rate of four miles an hour, maintained for sixteen or seventeen hours.

In later life Agassiz made to us some prophecies in which we may trace his enjoyment of the finest physical health and strength. Health and strength indeed belonged to everything which he said and did.

Among other things he said, twenty-five years ago, that the last years of our century — the twentieth — would see a population of a hundred million of people in the valleys of the upper Amazon. I like to keep in memory this brave prophecy because I am sure it will come true.

James Dwight Dana[7]

It was my privilege to know James
Dwight Dana intimately during my early
years. To boyhood's imagination his figure
typified the man of science; his life
personified the spirit of scientific discovery.
Wider acquaintance with the world has not
in any way dimmed the brightness of that
early impression.

The services of the geologist are to-day
recognized by everyone, and sought by all
who can afford them. If he would make a
voyage of exploration and discovery, the
resources of the world of finance are placed
at his disposal. No such aids were given two
generations ago.

---

[7] By Arthur T. Hadley

In Dana's journeyings he had to surmount hardship and peril, and to meet the coldness of those who knew not the value of the quest which he pursued. He and his contemporaries were like the knights errant of chivalry, devoting their lives to an ideal. They were men of faith, who combined the spirit of the missionary and the inspiration of the poet with the clear vision of the observer.

The largeness of Dana's work was commensurate with the largeness of his inspiration. It fell to his lot not only to fill out many pages of the record of the building of the world, as written in the fossil life of America, but to show in important ways the methods by which that building was accomplished. His creative brain never rested content with mere description of facts. He had the more distinctively modern impulse to reconstruct the process by which those facts were brought to pass. From his observations of coral islands in the various stages of their growth he deduced a geologic principle of world-wide importance. It is this characteristic which makes the great modern German school of geologists headed by Suess look to Dana as their precursor, more than to any other man of his generation.

He was not content with the work of discovery alone. The teaching spirit was strong within him. The pioneers in science needed editors and expositors who should make their results known. In each of these capacities Dana's achievements were phenomenal. Of his work as an editor, he has left the files of The American Journal of Science as a monument. Of his work as an expositor those who have heard his lectures and attended his class-room exercises can speak with un- bounded enthusiasm. He was one of the rare men who by presence and voice and manner could bring the truths and ideals of science home even to those pupils with whom scientific study could never be more than an incident in their lives.

But above all his works and above all his qualities stands the figure of Dana himself — more than an explorer, more than a discoverer, more than a teacher; his countenance, as it were, illuminated by a touch of the light of a new day for which the world was being prepared.

His life was gentle; and the elements
So mixed in him that Nature might stand forth
And say to all the world, 'This was a man.'

Spencer Fullerton Baird[8]

The life, the character, the work of Spencer Fullerton Baird entitle him to recognition in any assemblage and on any occasion where honor is to be paid to those who have been their county's benefactors through illustrious achievements in science.

Developing a taste for scientific pursuits at a very early age, and confirmed in those pursuits through the influence of friendships with Agassiz, Audubon, Dana and other leading scientists of the time, Baird was selected as assistant secretary of the Smithsonian Institution when only twenty-seven years old, and there entered on a career devoted to the promotion, diffusion, and application of scientific knowledge

[8] By Dr. Hugh M. Smith

among men, and marked by dignity, sound judgment, fidelity to duty, versatility and general usefulness.

In the many phases of his intellectual development he resembled Franklin and Cope; in the multiplicity of his public duties and in the diversity of the scientific accomplishments in which he attained eminence he had few equals; in founding, organizing and simultaneously directing a number of great national scientific enterprises he was unique among those whose memory is here extolled today.

To render an adequate account of the branches of scientific endeavor in which he achieved prominence, benefited his own and future gen rations and added to his country's renown, one would need to be an ornithologist, a mammalologist, an ichthyologist, a herpetologist, an invertebrate zoologist, an anthropologist, a botanist, a geologist, a paleontologist, a deep-sea explorer, a fishery expert, a fish-culturist, an active administrator of scientific institutions, and an adviser of the federal government in scientific affairs; for Baird was all these and more.

We freely acknowledge today the debt that science owed Baird alive and now owes his memory, especially for his inestimable services as assistant secretary and later as secretary of the Smithsonian Institution, as director of the National Museum, and as head of the Commission of Fish and Fisheries. Among all the establishments with which he was connected, this last was preeminently and peculiarly his own. It was conceived by him and created for him, and it would almost appear that he was created for it, for certainly no other person of his day and generation was so admirably fitted for the task of organizing this bureau and of executing the duties that grew out of its functions as successively enlarged by congress. Insisting on scientific investigations and knowledge as the essential basis for all current and prospective utilitarian work, he drew around him a corps of eminent biologists and physicists; he established laboratories; he laid plans for the systematic study of our interior and coastal waters ; he had vessels built that were especially designed and equipped for exploration of the seas.

While he thus inaugurated operations which have been of lasting benefit to the fisheries, at the same time he became the foremost promoter and exponent of marine research, and the knowledge we today possess of oceanic biology and physics is directly or indirectly due to Baird more than to any other person. The rapid development of piscicultural science under his guidance gave to the United States the foremost place among the nations in maintaining and increasing the aquatic food supply by artificial means; and it was no perfunctory tribute when, in 1880, at the International Fishery Exhibition held in Berlin, Emperor William awarded the grand prize to Baird as ' the first fish-culturist in the world.'

The spirit of Baird influences the Bureau of Fisheries today, as it does all other institutions with which he was associated; and since his death, nearly twenty years ago, the good that has been accomplished in the interest of fish-culture and the fishing industry, and in the conduct and encouragement of scientific work, has been in consequence of the foundations he laid, the policy he enunciated and the example he set.

But conspicuous as were his services to science and mankind; faithful and unselfish as was his devotion to the executive responsibilities imposed on him; beautiful as was his personal character, I conceive that his most enduring fame may result from the enthusiasm with which he inspired others and the encouragement and opportunity that he afforded to all earnest workers.

The recipients of his aid can be numbered by hundreds, and many of them are today his worthy successors in various fields; and their places in turn will gradually be taken by a vast number of men and women who will perpetuate his memory by efficiently and reverently continuing his work.

This evidence of the donor's beneficence is a noble and impressive memorial of one who merited his country's profoundest gratitude; but the bust signifies something more, for it is a recognition of that zeal, fidelity, self-sacrifice, intelligence and strength in the American character so preeminently typified by Spencer Fullerton Baird.

Joseph Leidy[9]

Joseph Leidy was born in Philadelphia, there he passed his three score years and ten, and there he died. For forty-five years he was an officer of the Philadelphia Academy of Natural Science, and a professor in the University of Pennsylvania for forty years. His character was simple and earnest, and he had such a modest opinion of his talents and of his work that the honors and rewards that began to come to him in his younger days, from learned societies in all parts of the world, and continued to come for the rest of his life were an unfailing surprise to him.

His knowledge of anatomy, and zoology, and botany, and mineralogy was extensive

[9] By Pr. William K. Brooks

and accurate and at his ready command. Farmers and horticulturists came to him and learned how to check the ravages of destructive insects; physicians sent rare or new human parasites and were told their nature and habits and the best means of prevention; jewelers brought rare gems and learned their value. His comments, at the academy, on the recent additions to its collections, gave a most impressive illustration of his ready command of his vast store of natural knowledge.

Leidy wrote no books, in the popular meaning of the word. He undertook the solution of no fundamental problem of biology. There are few among his six hundred publications that would attract unscientific readers, or afford a paragraph for a newspaper. They are simple and lucid and to the point. Most of them are short, although he wrote several more exhaustive monographs. They cover a wide field, but most of them fall into a few groups. Many deal with the parasites of mammals — among them, one in which his discovery of Trichena in pork is recorded.

Two hundred and sixteen, or about a third of his publications, are on the extinct vertebrates of North America. His first paper on paleontology was published in 1846, and his last in 1888, as the subject occupied him for more than forty years. He laid, with the hand of master, the foundation for the paleontology of the reptiles and mammals of North America, and we know what a wonderful and instructive and world-renowned superstructure his successors have reared upon his foundation. It was this work that established his fame and brought him honors and rewards. They who hold it to be his best title to be enrolled among the pioneers of science in America are in the right in so far as the founder of a great department of knowledge is most deserving of commemoration; but I do not believe it was his most characteristic work.

I can mention but one of the results of his study of American fossils. He showed, in 1846, that this continent is the ancestral home of the horse, and he sketched, soon after, the outline of the story of its evolution which later workers have made so familiar.

More than half his papers are on a subject which seems to me to contain the lesson of his life. Like Gilbert White, he was a home-naturalist, devoted to the study of the natural objects that he found within walking-distance of his home, but he penetrated far deeper into the secrets of the living world about him than White did, finding new wonders in the simplest living being. In the intestine of the cockroach and in that of the white ant, he found wonderful forests of microscopic plants that were new to science, inhabited by minute animals of many new and strange forms. His beautifully illustrated memoir on A Flora and Fauna within Living Animals is one of the most remarkable works in the whole field of biological literature. Another memoir gives the results of his study of the anatomy of snails and slugs. The inhabitants of the streams and ponds in the vicinity of his home furnished an unfailing supply of material for re- search and discovery, and many of his publications are on aquatic animals. He finally became so much interested in the fresh-water rhizopods that he abandoned all other scientific work in order to devote all his attention to these animals. His results were published in the memoir on *The Fresh-water rhizopods of North America.* This is the most widely known of his works. It is, and must long be, the standard and classic upon its subject. I have no time to dwell upon his work as the naturalist of the home — his best and most characteristic work. Its lesson to later generations of naturalists seems to me to be that one may be useful to his fellowmen, and enjoy the keen pleasure of discovery, and come to honor and distinction, with- out visiting strange countries in search of rarities, without biological stations and marine laboratories, without the latest technical methods, without grants of money, and, above all, without undertaking to solve the riddles of the universe or resolving biology into physics and chemistry.

If one have the simple responsive mind of a child or of Leidy, he may, like Leidy, ' find tongues in trees, books in the running brooks, sermons in stones, and good in everything.'

Edward Drinker Cope[10]

In the beautiful marble portrait of Edward Drinker Cope, modeled by Mr. Couper and presented by President Jesup, you see the man of large brain, of keen eye, and of strong resolve, the ideal combination for a life of science, the man who scorns obstacles, who while battling with the present looks above and beyond. The portrait stands in its niche as a tribute to a great leader and founder of American paleontology, as an inspiration to young Americans. In unison with the other portraits its forcible words are : 'Go thou and do likewise.'

Cope, a Philadelphian, born July 28, 1840, passed away at the early age of fifty-

[10] By Pr. Henry F. Osborn

seven. Favored by heredity, through distinguished ancestry of Pennsylvania quakers, who bequeathed intellectual keenness and a constructive spirit. As a boy of eight entering a life of travel and observation, and with rare precocity giving promise of the finest qualities of his manhood. Of incessant activity of mind and body, tireless as an explorer, early discovering for himself that the greatest pleasure and stimulus of life is to penetrate the unknown in nature. In personal character fearless, independent, venturesome, militant, far less of a quaker in disposition than his Teutonic fellow citizen Leidy. Of enormous productiveness as an editor, conducting the American Naturalist for nineteen years, as a writer leaving a shelfful of twenty octavo and three great quarto volumes of original research. A man of fortitude, bearing material reverses with good cheer, because he lived in the world of ideas and to the very last moment of his life drew constant refreshment from the mysterious regions of the unexplored.

In every one of the five great lines of research into which he ventured, he reached the mountain peaks where exploration and discovery, guided by imagination and happy inspiration, gave his work a leader- ship. His studies among fishes alone would give him a chief rank among zoologists, yet among amphibians and reptiles there never has been a naturalist who has published so many papers as Professor Cope, while from 1868 until 1897, the year of his death, he was a tireless student and explorer of the mammals, living and extinct. Among animals of all these classes his generalizations marked new epochs. While far from infallible, his ideas acted as fertilizers on the minds of other men. As a paleontologist, enjoying with Leidy and Marsh that Arcadian period when all the wonders of our great west were new, from his elevation of knowledge which enabled him to survey the whole field, with keen eye he swooped down like an eagle upon the most important point.

In breadth, depth and range we see in Cope the very antithesis of the modern specialist, the last exponent of the race of the Buffon, Cuvier, Owen and Huxley type. Of ability, memory and courage sufficient to grasp the whole field of natural history. As comparative anatomist he ranks with Cuvier and Owen; as paleontologist with Owen, Marsh and Leidy — the other two founders of American paleontology; as natural philosopher less logical but more constructive than Huxley. America will produce men of as great, perhaps greater, genius, but Cope represents a type which is now extinct and never will be seen again.

www.ingramcontent.com/pod-product-compliance
Lightning Source LLC
Chambersburg PA
CBHW060414190526
45169CB00002B/898